A GEOLOGICAL REPORT

OF THE

COAL AND OIL LANDS

IN

PULASKI, WHITLEY, AND WAYNE COUNTIES,

KENTUCKY,

BELONGING TO THE

"CUMBERLAND RIVER COAL COMPANY OF KENTUCKY."

BY J. M. SAFFORD, A. M.,

PROFESSOR OF CHEMISTRY AND GEOLOGY IN CUMBERLAND UNIVERSITY, LEBANON, TENN.,
AND FORMERLY STATE GEOLOGIST OF TENNESSEE.

LOUISVILLE, KY:
PRINTED BY JOHN P. MORTON & CO., NO. 425 MAIN STREET.
1865.

REPORT.

In accordance with your instructions, I have made a reconnoisance of your lands in Pulaski and Whitley counties, Kentucky. These lands, the location and extent of which may be seen upon the map in course of preparation by your Company, form, in the aggregate, a very large territory—so large that a thorough examination of the whole area would require much more time than I have been able to bestow upon it. Large portions, however, have been examined, and enough done to demonstrate to my own mind the great value of the entire property. It is a great coal field. Its strata abound in iron ore, and its forests in excellent timber. Its large streams present mill-sites and water power rarely excelled. Its high, flat, wooded plains have been (and may be again, now that peace is at hand,) the favorite summer pasture-grounds of numerous herdsmen. Its cool, salubrious climate and pure waters, together with its varied scenery, make it a desirable summer retreat, and will make it, when its hidden wealth is in the height of development, a region of delightful homes and pleasant villages.

1st. General Topographical Features.

The region including your property is a portion of the great Cumberland Plateau of Kentucky and Tennessee, the general elevation of which is here not far from 2,000 feet above the sea. The region is thus, in general, a table-land or plateau, with a surface comparatively flat or gently rolling. The traveler may pass, on some parts of it, for miles through the "flat woods," without meeting with any deep valleys. This especially is the case, when his route lies between the larger water courses. But yet, while this is its general character, it is not an unbroken table-land. It is deeply cut by the Cumberland river, and by the main tributaries of the Cumberland, the Rockcastle river, and the Big South Fork, as well as by the smaller tributaries of all these streams. These streams flow, in the main, from the circumference of the region to a common central district, and finally concentrate in the Cumberland river. The whole plateau is thus beautifully intersected and laid open by a fan-like system of narrow valleys. Point Burnside may be regarded as the radiating point of the whole system. From Burnside we have access to the great plateau in a southerly direction up the valley of the Big South Fork, or in an easterly direction up the Cumberland. A few miles above Burnside the main valley of the Cumberland is fairly within the plateau. By this system of valleys the nearly horizontal strata of the plateau are well exposed and their included beds of coal, ore, &c., made easily accessible.

The three rivers mentioned—the Cumberland and its tributaries, the Big South Fork, and the Rockcastle, divide the region including your lands into three distinct areas. The first lies on the North side of the Cumberland, between Buck Creek and the Rockcastle; the second, north and east of the Cumberland and east of the Rockcastle; the third, between the Cumberland and the Big South Fork. The first area, that between Buck Creek and the Rockcastle, is much cut up by the valleys of small streams. It is in fact a bed of broad, flat-topped ridges, which have in general an elevation of from 400 to 500 feet above the

Cumberland river. This bed of ridges consists of a main ridge dividing the waters of the Rockcastle and the Cumberland from those of Buck Creek, from which numerous spurs, separated by small branches, run off to the main water courses. On the east side of the Rockcastle, in the second area, the plateau is much cut into such ridges as we have just described, and having about the same elevation; but further back and above the Falls of the Cumberland, it is not as broken. Between the Cumberland and the South Fork, in the third area, we have a main dividing flat-topped ridge with spurs, much as in the first area. The heavy ridges fall away towards Burnside Point, but begin to press in abruptly on the river, five or six miles above. These ridges have about the same elevation as those in the first area.

The character of the river and creek valleys may be in part inferred from what has already been said. They are generally narrow. The ridges crowd closely into the streams, often leaving a narrow strip of bottom land, but sometimes none. Along, on both sides of the valleys, the slopes of the ridges are capped by great overhanging cliffs of sandstone, or conglomerate. These cliffs are most prominent on the Cumberland and Rockcastle, above the mouth of the latter. Below this point the high cliffs recede more or less from the river, and limestone bluffs, rising often from the water's edge, begin, as we descend the river, to be a conspicuous feature along its banks. Above the Falls of the Cumberland the river may be said to be almost upon the plateau; at the Falls it pitches into its narrow, deep, cliff-bound valley, which is here much more contracted than at the mouth of the Rockcastle. The valley of the Big South Fork presents very much the same features as that of the Cumberland. Starting from the mouth of the river and ascending, the limestone bluffs, rising almost out of the water, are conspicuous; several miles up, these become less prominent, and finally disappear, the sandstone cliffs at the same time towering above the contracted valley.

2d. General Geological Character.

But two great geological formations are seen within the limits of your lands. They are:—

1st. *The Coal Measures.* By this is meant the whole series in which are contained the beds of coal. The maximum thickness of this formation is, in the more southern and eastern portions of your lands, not less than 600 feet. This series rests upon—

2d. *The Mountain or Sub-Carboniferous Limestone.* This is a heavy series of mostly limestone strata. Its upper part is often made up of variegated shales, with beds of impure limestone, ("bastard" limestone). The entire thickness of this formation, on the Cumberland, above the mouth of the Big South Fork, may be placed at about 300 feet. It increases in thickness in a southerly direction.

Approaching your property from the west we find the bed of the Cumberland and the beds of the larger creeks cut out of the upper part of the second formation—the mountain limestone. Just within the limits of the property, the limestone is found, in the bluffs and upon the hillsides, as high as 150 feet above high-water mark of the Cumberland. This, then, is about the general elevation of the top of the limestone, along the extreme north-western limits of your lands. Now from this limit the whole formation dips south easterly, with such an inclination as to cause its topmost strata to pass below the bed of the river, two or three miles above the mouth of the Rockcastle, and to disappear. The last of the limestone is a little below the old Hudson mines. Going up the Big South Fork the limestone sinks in much the same way. At the old Beaty salt well its top is seen below high-water mark.

The coal measures, resting as they do upon the limestone, will of course, along the north-western limits of your property, have their bottom strata raised as high as the top of the limestone—about 150 feet in general above high-water. Here the base of the ridges is limestone, their middle and upper parts coal measures. Passing, however, to the east or south-east, the limestone strata, as I have said, finally sink beneath the river, and the whole country—valley and high-land—becomes coal measures. All your lands, however, belong properly to the coal measures, as the limestone beds of the streams in the western portion of the region occupy comparatively but little area.

The strata of the coal measures, as they occur upon your property, are divisible into two parts, the upper and lower coal series, and it is the lower series only that has been anything like fully developed. The upper series lies in the eastern and southern portions of your property, and doubtless contains beds of coal and iron ore of great value not anticipated by your company. But more of this hereafter.

The lower coal series extends over the whole of your territory, with simply the exception of the narrow limestone beds of the streams already referred to.

The following general section of this series, made up from many observations, will aid in giving a clearer idea of the succession of strata and of the geological structure of the country.

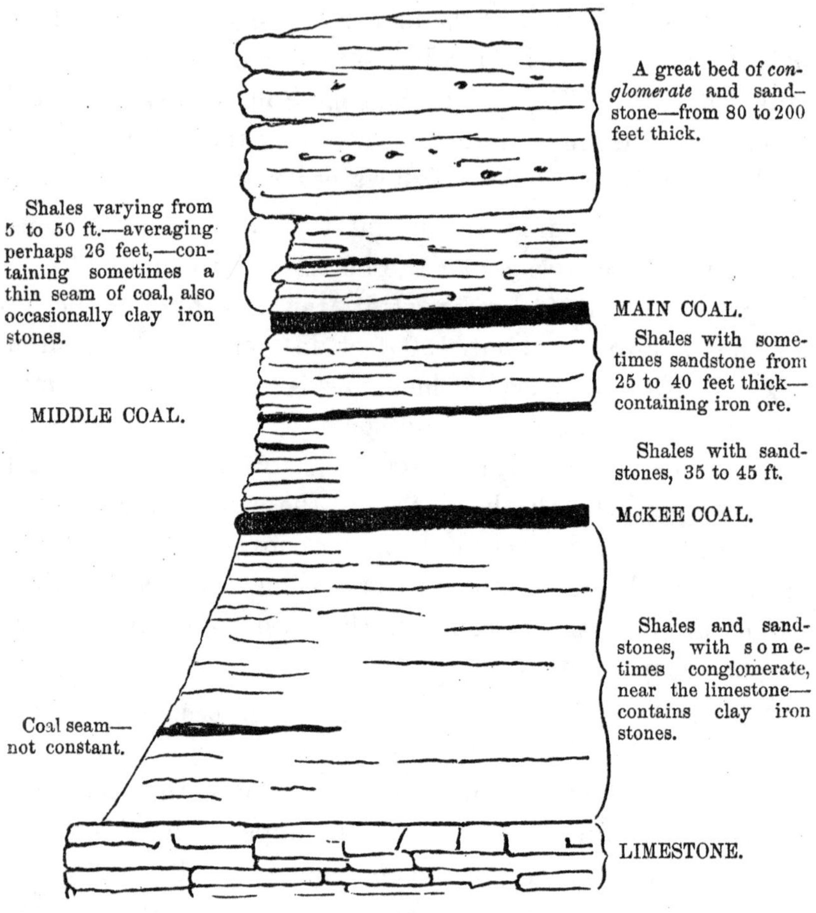

The conglomerate is an important rock in this region. It caps the ridges and forms the overhanging cliffs, which are so conspicuous at many points. Its cliffs line the valley of the Cumberland up to the Falls, at which point the river descends from above it. It forms, in general, the topmost rock until we reach the eastern and southern portions of your property, where it is often covered by the upper coal series.

3D. COAL.

There are five beds, and sometimes six beds of Coal in this region. Of these, two beds are workable, the *Main Coal*, and what is best known as the *McKee bed.*

The *Main Coal* has been opened at numerous points, and on both sides of the river. Within the region west of Rockcastle river, and north of the Cumberland, there are many banks. Here the plateau, as we have stated, is so cut by ravines as to be, in fact, a bed of ridges; upon the sides of these ridges the coal strata crop out, and are easily accessible. Tramways have been built, (the position of which may be seen upon the map,) and others may be built, ascending from the river, up the ravines, by easy grades to the main coal. It is unnecessary to enumerate these banks, as their places will be designated on the map. With the exception of those on Big Lick Creek, they are in the main coal bed. They all have the same general character, and it is from them that most of the coal, taken to Nashville, and down the Cumberland, has been derived. The bed of coal, so well exposed by these openings, at many and distant points, is a most valuable and excellent one. It varies in thickness from 36 to 60 inches solid coal, averaging nearly, or quite, in this district, about 50 inches. It is a compact, cubic, free-burning, moderately bituminous coal, and has been, of all the coals taken to Nashville, the most desired for common purposes. Of its superior quality the older citizens of Nashville, who are acquainted with its use, will testify.

Resting upon this main coal, is almost always a stratum of black, slaty, cannel coal, from a few inches to nearly two feet in thickness, which, doubtless, may be turned to some practical account, especially as the mining of it may be carried on in connection with that of the coal below, thus costing comparatively little. Heretofore, like the slack from the mines it has generally been thrown away. It forms, however, very often a good roof and is not taken out. This slaty cannel burns freely, and has been used at the mines as a fuel. It might be used also in producing petroleum.

In this portion of your lands alone, the amount of coal in this main bed is very great, and it is, of itself, a mine of wealth. It is found in all the ridges of this region, on the south side of the river. Opposite the mouth of Addison's branch, at the old Huling banks, it also presents a fine exposure. Not much has been taken out, but the banks are well opened, the coal presenting a face of from 55 to 60 inches, with a foot and a half of the slaty cannel above it.

The horizontal extent of the bed is great, varying more or less in thickness. It is found at numerous points on the ridge sides bounding the valley of the Big South Fork, towards the southern boundary of the State. It is found near and above the mouth of the Rockcastle. Further up, in Whitley County, it remains in good part undeveloped. That it exists, however, there can be no reasonable doubt. It will, however, beyond the Falls of the Cumberland, run, with the conglomerate above it, under the beds of the streams.

The next most important coal bed is that known as the *McKee bed.* This lies from 60 to 90 feet below the last, and with it pervades the same country. The McKee bed has been opened and worked at several different points. One of these is on Big Lick creek, about one and a half miles from the Cumberland river. The bank was opened several years since, by Mr. Alexander McKee. Owing to the breaking out of the rebellion, it has not been worked to any great extent. A railroad track was constructed from this bank down to the river. The coal is here from four to nearly five feet in thickness. It has a seam of clay running through it, about a foot from the top, which separates it into two parts. The bed was also worked at a point about two miles distant, near the river, a little above the mouth of Big Lick. It was here about three feet thick. Some coal was taken from it also, at one point on Addison's branch. It has, however, been most extensively worked at the old Hudson mines, on the Cumberland, three or four miles above the mouth of Rockcastle river. Much coal has been taken from this point to Nashville. It is here from three to four feet thick, averaging about three and a half feet, and owing to the general dip of the rocks, is but just above high-water mark. The same bed of coal is seen at various points, of good thickness, along the Big South Fork. It has been opened at the old Beaty salt well, but the bank has since fallen in. It was reported to be five feet thick. I had its out-crop exposed, and measured three and a half feet clear good coal, and think it may reach, when opened, nearly or quite the thickness reported. The bed is thus one of great extent; its thickness, more variable than the main coal above, ranging from five feet down to three feet, and at some points still less. It is a highly valuable bed, and in position, if not in quality of coal, has the advantage of the upper bed.

At the McKee bank, I have said it has a clay parting. This feature is represented at Hudson banks by a slate parting, about an inch thick, and at other points in the vicinity of the Cumberland the bed may be recognized by this feature. The coal of this bed on the Cumberland and Big Lick has a laminated structure, and inclines to break in flat pieces. At other points its blocks are more cubical.

It is a free burning coal, rather more so than that of the main bed, and, at the Hudson banks, remarkably free from pyrites, which for some purposes, would make it greatly desired. The remaining coal beds are not generally of workable thickness; they vary from mere thin seams to two or perhaps three feet. At some points they are thick enough to work, but they cannot be depended upon.

The McKee bed, at the Hudson mines, is so near the river, being but little above high-water mark, that it can be immediatly transferred to the boats from the mine. Going down the river, the coal beds rise in the ridges. Below the mouth of Rockcastle, the coal is conveyed from the banks to the bluffs along the river by railways. From these bluffs it is transferred to the boats. Nature has presented in the ravines, cutting the coal rocks in the limestone ledges along the water's edge, and in the eddies of the river, excellent facilities for moving and shipping the coal.

Much improvement can be made upon the methods heretofore used. For many years before the rebellion, coal was shipped down the river. Nashville is not the only market. In fact the market is the whole country below Nashville, both on the Cumberland and the Mississippi rivers, as far down as New Orleans. Once at Nashville, the whole country below is open. With moderate improvements in the channel of the Cumberland at the shoals above the mouth of the Big South Fork, boats could descend as frequently, perhaps, as the interests of the company might require.

The coal, heretofore, has been shipped in flat-bottomed boats, built near the banks. Timber is abundant for this purpose.

I have said that the coal rocks of your lands are divided into two series, an upper and a lower. What has been said has been confined to the lower series. The upper series has not been developed, but there is good reason to think it will furnish an excellent bed of coal. I have given, on a previous page, a section of the lower series. This is capped by the conglomerate. It is upon this conglomerate that the upper series rests. Some of its shales are found at a few points on the ridges west of the Rockcastle. East of this, however, and on the plateau, about and above the Falls, there are ridges of considerable elevation, which contain beds of shales and sandstone. Further eastward and southward the series must be of considerable thickness, and doubtless reaches three or four hundred feet. In Tennessee, at numerous points, there is an excellent bed of coal, from four to six feet in thickness, at a distance above the conglomerate, varying from fifty to a hundred and forty feet. Upon your property there is sufficient thickness of strata to include this bed, and an examination will doubtless bring it to light. In fact the croppings of a bed of coal in the shales above the conglomerate are known not far from the Hudson banks. Whether this is the bed in question or not I cannot say. Be that as it may, further research will doubtless be rewarded by the development of a good bed of coal in this part of the coal series.

4TH. EXTENT AND QUANTITY OF COAL.

As to quantity the coal upon your lands may be said to be practically inexhaustible. Many generations will not see its end. It must be borne in mind that coal strata very generally spread out, in a region having the geological structure of yours, with their accompanying beds of shales and sandstones, for miles,—sometimes for hundreds of miles. The two strata of the lower coal series, which have been designated respectively as the "main coal," and the "McKee coal," and already traced out and known for miles, are to be confidently looked for, in their proper places, *over the whole extent of your territory*, excepting only the narrow valleys and ravines, in its western and north-western portions. Their thickness will not be constant, but it will be the exception not to find them. Even the smallest strata may be looked for. And, as I have said, in the eastern and southern portions of the property, where the upper series of the coal rocks occur, we have in addition to those mentioned, and above them, in all probability, another valuable bed of coal.

If we take, and this doubtless is a minimum estimate, the coal strata of your lands to be equal, collectively, to a single uniform stratum, five feet thick, and co-extensive with the lands, excepting the ravines mentioned above, a simple calculation will show the vast amount of coal upon your property. In this case every acre contains (allowing 2,688 cubic inches to the bushel,) 140,013 bushels of solid coal, which, when mined and thrown into boats or coal-yards, and having the form in which it is ordinarily sold, would measure at least 150,000 bushels. Thus, then, 100,000 acres contain 15,000,000,000 bushels, worth, at a cent a bushel, $150,000,000.

5TH. IRON ORE AND TIMBER.

The iron ore upon your property is, in the aggregate very great. Nearly all the beds of shales contain it. It occurs at various levels in uniform nodules and masses, and sometimes in layers. Its development has not received as much attention as that of the coal, and it is mostly by its accidental out-crops that it is known.

It is mostly in the form of clay iron stones, the form most usually found in the coal measures. Some of it was worked at an old forge, now gone, formerly in operation on Buck Creek. The shales above the conglomerate, may, upon examination, also be found to contain it. The whole region is well worthy the attention of iron men. As to what might be the best point for the location of an iron establishment, will have to be determined by further examinations. Timber for charcoal, is abundant, and the coal beds can of course supply coal, if required, all being upon the same ground.

As to timber, a few words will be sufficient. There is no lack of timber upon the whole of your tract. It abounds everywhere, and is often of most excellent quality. Along the bluffs of the river, especially where the limestone appears, the growth is Beech, Maple, Poplar, Linn, Buckeye, Ash, White Oak, Hickory, Walnut, &c. On the slopes and tops of the ridges, it is white, red, black and chestnut Oaks, Poplar, Yellow Pine and Chestnut. These trees are often of very large size, tall and straight, suitable for any of the purposes for which such timber may be used. At numerous points the Chestnut Oaks are valuable for the bark they might afford for tanning purposes.

6TH. PETROLEUM.

Without doubt your territory is within the petroleum district of Kentucky, and, for good reasons, certain portions are especially promising. These are made so by their general geological and topographical character. One of these regions is in that part of the valley of the Big South Fork in which are located the old Beaty oil and salt wells, and to which your lands extend. Another is the valley of the Cumberland, from the mouth of the Rockcastle up by the old Hudson mines. In both regions, as stated on a previous page, the coal measures, by the dip of the rocks, are brought down to the level of the river.

But first let us notice the formations concerned. I have already mentioned the two great formations seen at the surface upon your lands; those which are the main source of petroleum lie below these. The complete series is as follows:

1*st. Coal Measures.* See page 6.

2*d. The Mountain Limestone.* See page 6.

3*d. The Knobstone Formation.* This lies next below No. 2, or the mountain limestone. It does not appear at the surface in much force upon your land, but, in consequence of the rise in the formations in a westerly and north-westerly direction, it is well seen in the hill sides along the Cumberland, below Point Burnside. A good section of the strata making up this formation is presented in the hill sides along the river, on the Somerset and Monticello road. The formation is from 400 to 500 feet thick, and consists, in the main, of impure, fine sandy and clayey layers, more or less calcareous. It is generally compact, of a light bluish color when freshly quarried, but weathering easily into shales, which are often fetid. It has been called properly a *mud-rock.* Occasionally harder interstratified beds of limestone occur in this formation. In some regions, especially southward in Tennessee, it abounds in heavy flinty layers.

4*th. The Black Slate.* This easily recognized formation of Devonian age in geology is from 40 to 60 feet thick. It is remarkable for its persistency, extending, as it does, from Alabama, through Tennessee and Kentucky, into Ohio, and northward. It is a black, fetid slate or shale, and is, doubtless, in connection with the fetid shales above (those of form 3), the principal source of oil in southern Kentucky.

Such are the formations which interest us at present. The first two are seen at the surface, and the others can be found anywhere upon your property by boring.

Now these formations are so disposed as to throw the

region of the Beaty wells within a natural geological depression or basin. Their rocky strata dip from the northwest and south towards the region of the wells, bringing the bottom of the coal measures down to the river, thus forming a natural reservoir—the basin mentioned. The black slate must be not far from 800 feet below the surface of the Big South Fork at the oil well, while it is but little below the surface at the mouth of the stream on the Cumberland. Now, allowing for the fall in the Big South Fork, the dip southward towards the wells must amount to at least 600 feet. This gives an idea of the depression of the basin.

We have here evidently the conditions of a valuable oil region. Here are at the bottom, and on the gradually ascending sides of a natural and extensive depression, the oil producing rocks; below these are impervious strata to retain the products, while above lies a formation (No. 2) in whose porous strata, and in the caverns and fissures for which it is noted, these products—the yield of ages—may be stored.

Moreover, direct evidence of the presence of petroleum in this region is furnished by the products of the wells bored for salt. One especially, known as the "Oil Well," and already referred to, is famous. Its history is given in the following letter from the Hon. E. L. Van Winkle, Secretary of State for the State of Kentucky.

FRANKFORT, KY., June 28th, 1865.

S. S. BUSH, ESQ.:

Dear Sir—I answer your inquiries in regard to the Beaty Oil Well as follows :

In 1825, or perhaps earlier, Conn, Beaty, and Fulton, of Abingdon, Va., opened three or four salt wells upon the Big South Fork of the Cumberland River—the one in question being one of the number. At the depth of 110 feet they struck oil, which flowed so freely, and so interfered with their work and machinery that the well was abandoned, and one sunk some two miles below. This latter well proved a complete success, as superior salt water was found at near 300 feet.

Some years after the oil well had been abandoned, one Marcus Huling concluded to try his fortune upon oil, and procured barrels and gathered a boat load of the oil from the well, and shipped it down the river. By some sort of mismanagement Huling lost his cargo a few miles below the well. This misfortune chilled the ardor of this enterprising oil pioneer, and from that time he ceased work.

In 1840 one Dr. James B. White, of Danville, employed hands, and sent them to the well for the purpose of gathering oil, and, with the aid of a horse pump, he was able to get about three barrels per day of the oil, which he sold in Cincinnati and other places for medical purposes.

That whole region abounds with surface indications of oil. Oil was found in every well that Beaty & Co. sunk upon the Big Fork, but at that date it was not sought after or considered valuable.

Yours truly,

E. L. VAN WINKLE.

In regard to the region above the mouth of the Rockcastle, the same formations and very nearly the same conditions exist as upon the Big South Fork. The easterly dip of the formations, bringing the coal measures down to the river, makes a trough-like depression in which petroleum could and most likely has accumulated.

Altogether, *there are no regions in southern Kentucky the general features of which are more promising, and in which the enterprising well-borer will be more likely to meet with success.*

In addition to the resources of your property already mentioned, salt water ought not to be forgotten. I do not propose to dwell upon this. It is enough to say that salt water has been found at numerous points in this part of Kentucky, and, as in the case of petroleum so with salt water, no localities in this part of the State are more favorable for obtaining it than many upon your lands.

Altogether, your whole territory is one of great promise. In addition to its known resources, new developments will be made, now unanticipated. With proper management, it will be a source of wealth to its possessors, and of comfort and support to hundreds of working-men.

JAS. M. SAFFORD.
Formerly Geol. of Tenn.

Statements of Messrs. Jasper and Hudson, with reference to cost of mining and shipping coal.

Before the war, boats cost, when 25 feet wide, $2 per foot of length. Thus, a boat of this width and 100 feet long would have cost $200. When over 25 feet in width, each additional foot, taken as many times as 25 would go in the length, was added to the length in the calculation, which was then multiplied by 2 to obtain the cost. Thus a boat 30 feet wide and 100 long, would cost as follows:

$$\begin{array}{r} 100 \\ 5\times4=20 \\ \hline 120 \\ 2 \\ \hline 240 \end{array}$$

Or, 240 dollars.

The coal, mined and delivered at the mouth of the bank, cost per bushel from	2 to 3	cts.
When placed on the river bank, per bushel	4	"
Loading, per bushel from	¾ to 1	"
Whole cost at Nashville, in the boat, per bushel	11	"
To unload, and place in yard, per bushel	2	"
Entire cost, per bushel	13	"
Wholesale price at Nashville, from	13 to 18	"
Retail price at Nashville, from	20 to 30	"

I will add that by proper arrangements, and a better system, the cost would be much reduced. Moreover, the slack at the mines could be converted into coke and saved; it is now all thrown away. There is every way much room for improvement.

The boats, when unloaded, are sold, I believe, for about forty or fifty dollars.

J. M. SAFFORD.

Statement about the coal from the banks of the upper Cumberland—the coal beds being in the lands of the Cumberland River Coal Company, of Kentucky.

The undersigned citizens of Nashville state that the city has been chiefly supplied with coal, for the last twenty years, more especially before the war, from the Addison branch, and other upper Cumberland coal banks. We state further, that the said coal is the best and the most desirable for common purposes that is brought to this market, always commanding a more ready sale than other coals.

S. M. SCOTT,
Formerly of the City Hotel.

J. KIRKMAN,
President Union Bank.

JOS. W. ALLEN,
Cashier Union Bank.

A. J. DUNCAN,
Pres. "Bank of the Union."

R. C. M'NAIRY,
Sec'y Com. Insurance Co.

W. T. BERRY,
S. J. CARTER,
Proprietors St. Cloud Hotel.

SAM'L E. HARE,
Formerly Commercial Hotel.

STATE OF KENTUCKY, EXECUTIVE DEPARTMENT.
FRANKFORT, June 6th, 1865.

MR. S. S. BUSH, LOUISVILLE, KY.:

Dear Sir—I have, for more than twenty-five years, been thoroughly conversant with that portion of the State in which your coal lands are situate, being a native of that part of the State of Kentucky. The coal upon those lands is a superior bituminous coal, with occasional veins of cannel coal. The coal beds are easily worked, and the facilities of water transportation good, and capable, at comparatively small expense, of being made equal, if not superior, to any west of the Alleghanies.

Truly yours,

THOMAS E. BRAMLETTE,

THE COMMONWEALTH OF KENTUCKY,
OFFICE OF SECRETARY OF STATE.

FRANKFORT, June 5th, 1865.

S. S. BUSH, ESQ., LOUISVILLE, KY.:

Dear Sir—Your map of the lands which I purchased for you of Wait, Hudson, and the "Nashville Coal Company," is before me, and I regard it as nearly accurate.

The lands represented as lying between the surveys marked respectively nine and fourteen should have been included within the boundary of those surveys.

The titles of these lands were carefully examined before I bought, and I append hereto a tabular statement which can be relied upon.

My long residence in the part of the State in which your lands are situated had enabled me to acquaint myself thoroughly with the property, as well as with the history of the titles, and consequently I had but little difficulty in selecting the best lands to which the titles were unquestionable. The Hudson banks or mines, it will be observed, are immediately upon the bank of the Cumberland River, and the loading water at these mines is not excelled by any upon that or any other stream: persons who have worked these mines informed me that boats could be loaded there *at any season* in the year.

The loading water at the terminations of the railroads is excellent, and is equally as good at a number of places along the river opposite the lands—say at mouth of Mill Creek, Beaver Creek, and opposite the mouth of Rockcastle River, and for miles above the latter point.

The 2,000 acre survey, numbered 4 upon the map, embraces in its boundaries near ten thousand acres of land, most of which was appropriated long before the date of this survey; the survey will, however, hold good for some three thousand acres on the north side thereof. All below or next to the river was taken by smaller and older surveys. All of those marked by black and red lines within the boundary of the 2,000 acre survey were bought of Wait and the Nashville Coal Company, the titles of which are unquestionable.

In addition to these surveys, you get of Wait and the Nashville Company a number of tracts lying upon Little Lick and Addison's Branch, which are not represented upon the map; these latter tracts are solid coal, and lay contiguous to the railroads on each branch, and have been two of the main points from which Wait has been mining.

The Iron railroad is embraced in the purchase from the Nashville Company, and leads to a solid coal region, some four thousand acres of which you acquired by the purchases.

The 8,000 acre survey, whose boundary is represented by blue lines upon the map, embraces some thirty thousand acres, and will hold good for at least twenty thousand acres, all solid coal, and includes, as you will observe, a number of older surveys, which you also own. I know of but one survey of 1,000 acres within the boundary anywhere near the river that is not owned by you.

The large surveys, Nos. 1 and 2, embrace some two hundred thousand acres of land, and inside thereof are many older surveys belonging to other persons; nevertheless, your purchase will hold some forty thousand acres within those boundaries, and, as a general thing, you will hold all the coal and iron lying anywhere near the river, the settlers having appropriated, as a general thing, the creek valleys only, for the purpose of getting farming lands. It follows that these large inclusive surveys embrace by far the greater portion of the mineral wealth within the boundaries of the same. These surveys are marked by yellow and brown lines upon the map.

Taken altogether you have substantially monopolized all that great coal region upon the main Cumberland.

The Phillips' lands, upon the Big South Fork, consisting of two tracts—500 and 2,278 acres—are conveniently located for approach to the South Fork, and lands lying contiguous have been mined with success, furnishing the same quality of coal as that found upon the main Cumberland, and can be transported or shipped with equal facility.

You will observe that as to some of the lands your vendor Wait holds equitable titles; that is, title bonds, most of which can be closed by deed at any time.

Yours truly,

E. L. VAN WINKLE.

CHARTER.

AN ACT TO INCORPORATE THE CUMBERLAND RIVER COAL COMPANY OF KENTUCKY.

Be it enacted by the General Assembly of the Commonwealth of Kentucky.

§ 1. That L. M. Flournoy, S. S. Bush, J. M. Bryant, and Moses Brown, and their successors, be, and they are hereby, created a body corporate and politic, by the name of the Cumberland River Coal Company of Kentucky, for the term of thirty years, with all the power and authority to corporations for the purposes hereinafter mentioned.

§ 2. The corporation is hereby authorized and empowered to purchase and hold lands in fee simple and by lease, for mining for coal, and boring for petroleum, and other oils and minerals, and to refine and vend the same.

§ 3. The capital stock of the Cumberland River Coal Company of Kentucky shall be one million dollars ($1,000,000), and shall be divided into shares of not less than five nor more than one hundred dollars each, and may be issued and transferred in such manner, and upon such conditions, as the board of directors of said corporation may direct.

§ 4. The affairs of said company shall be managed by four directors, one of whom shall be president, all of whom shall be stockholders in said corporation. The first board of directors shall consist of L. M. Flournoy, S. S. Bush, J. M. Bryant, and Moses Brown, who shall continue in office until their successors are elected; if any of the above named directors should decline or refuse to act, a majority of the others shall fill the vacancy by appointing some one else. They may adopt such by-laws and rules for the government of the corporation and management of its affairs and business as they deem proper, not inconsistent with the constitution and laws of the State. The said corporators, or any of them, may open books of subscription and receive subscription to the capital stock of said company herein incorporated; and books of subscription may be opened and subscriptions received, at such times and places, and upon such notices thereof, as any three of the said incorporators may deem right and proper.

§ 5. The said Cumberland River Coal Company of Kentucky shall not own in fee simple and by lease lands exceeding in value one million dollars, the capital stock of the company. Whenever one hundred thousand dollars of the capital stock of the company is subscribed, and ten per cent. thereof paid in, notice shall be given of the time and place of election of a new board of directors, who shall hold office for one year, and as provided herein. The board of directors of said corporation may fill all vacancies occasioned by death or resignation; are authorized and may make such calls of payment of stock as they deem proper; not exceeding twenty per cent. for every thirty days. They may keep their office at such place or places as they deem to the interest of the corporation. They may appoint a secretary, treasurer, superintendent, and other officers as they may deem necessary, with such compensation for services as they may fix, and by their laws regulate and fix the mode of keeping their books, as may be deemed necessary.

§ 6. This act shall take effect from its passage.

H. TAYLOR,
Speaker of House of Representatives.

RICHARD T. JACOB,
Speaker of the Senate.

By the Governor,
E. L. VAN WINKLE,
Secretary of State.

Approved Feb. 23, 1865.

THOS. E. BRAMLETTE,
Governor of Kentucky.

COMMONWEALTH OF KENTUCKY,
OFFICE OF SECRETARY OF STATE.

I, E. L. Vanwinkle, Secretary of State, and keeper of the archives thereof, do hereby certify that the foregoing copy of an act, approved Feb. 23, 1865, is a true and correct copy from the original enrolled law on file in this office.

In testimony whereof, I have hereunto set my hand and affixed my official seal.

Done at Frankfort, this 5th day of June, A. D. 1865.

E. L. VAN WINKLE,
Secretary of State.

www.ingramcontent.com/pod-product-compliance
Lightning Source LLC
LaVergne TN
LVHW011140110826
845150LV00008B/2426

* 9 7 8 1 4 1 8 1 9 3 3 4 8 *